Mobil New Zealand Nature Series
NEW ZEALAND NATIVE TREES

D0499464

Mobil New Zealand Nature Series
NEW ZEALAND NATIVE TREES I

Nancy M. Adams

REED

Published by Reed Books, a division of REED Publishing
(NZ) Ltd, 39 Rawene Road, Birkenhead, Auckland.
Associated companies, branches and representatives
throughout the world.

ISBN 0 7900 0139 X

© 1967 Nancy M. Adams
First published 1967
Reprinted 1968, 1969, 1971, 1972, 1973, 1974, 1977, 1980,
          1981 1985, 1987, 1989, 1991. 1993. 1994.

Printed in Singapore

FOREWORD

Few peoples have come to hold their native vegetation in such esteem as have New Zealanders; yet few peoples can have destroyed their native flora so vigorously and rapidly in converting what was once richly forested land to agricultural uses. Fortunately, even while this was happening to New Zealand, legislators had the good sense to reserve fairly large areas as National Parks, scenic and other reserves, and State forests. Most of these parks, reserves, and forests serve the essential purpose of preventing erosion and helping to absorb some of the heavy rainfall of the mountain lands, so reducing the intensity of flooding.

Many New Zealanders enjoy recreation and sport in these places and take pride in knowing something of the plants that make up the natural vegetation. For such people, Nancy Adams provides here, with accuracy and artistry, the details of some of the commoner trees.

A. L. POOLE
Former Director-General of Forests
New Zealand

CONTENTS

INTRODUCTION

FEW PLACES IN NEW ZEALAND are far from "the bush". In most of these areas of native forest, some of which cover many square miles, the types of forest that once covered the country may still be seen in something like their primitive state.

In the simplest terms these forests are classified into "podocarp-mixed broadleaf", "kauri", or "beech" forests, the names indicating the types of trees that are either the most conspicuous or the most abundant in their composition.

Podocarp-mixed broadleaf forests are those in which there are numerous tall trees belonging to the family *Podocarpaceae* i. e. rimu, kahikatea, matai, miro, silver pine and totara. They are all softwood timber trees, some with the alternative names of red, white, black and silver pine. Growing with them are many kinds of "broadleaf" hardwood trees such as kamahi, tawa, kohekohe, rata, and many others that form a leafy canopy above which stand the tops of the tall podocarps. Also included under this heading are: swamp forests with dense stands of kahikatea; forests of some dry areas where totara is the most abundant tree; high-altitude forest with kaikawaka (cedar); and coastal forest with many "broadleaf" species but few podocarps.

Apart from the magnificent trees in this type of forest there are many shrubs, tree-ferns, lianes and an abundance of small ferns, mosses, and lichens. Few trees are without epiphytes, which are perching plants that use the host tree for a place to grow upon but not for nourishment, as do parasites such as mistletoe. The epiphytes are of many kinds ranging from tree species—rata, puka, and others that germinate in pockets of humus high in the branches of large trees—to shrubs, astelias, orchids, and ferns. In forest where the rainfall is high the epiphytes may be so numerous as to make the identification of their host tree difficult. Mossy growths cover the trunks and paint-like lichens disguise the bark. When the trees grow too high for flowers and fruit to be gathered, an examination of fallen leaves, petals, and fruits will give an indication of the trees in the vicinity.

Kauri forest of the north of the North Island is similar to the podocarp-mixed broadleaf forest, but with the kauri as a frequent tree. Broadleaf trees are abundant in such forests, together with other trees and shrubs that do not grow further south than the kauri.

The exterior view of kauri forest is characterised by the tall kauris that stand well above the surrounding trees. High in their branches are huge clumps of perching plants. Within the forest the undergrowth is dense with "cutty grass" scrambling kiekie, shrubs and ferns. Tall nikau palms and tree-ferns reach up to the forest canopy, and taraire, tawa, toatoa, and towai are common trees. The enormous trunks of the kauri are covered with flaking bark that piles in mounds encircling the bases of the trunks. Flaking and peeling bark discourages the growth of epiphytes on the trunks of kauri and other trees such as miro, matai, rimu and totara.

Beech forests are those in which one or more species of southern beech grow in quantity. Except in mountain regions or in very dry beech forest, podocarps and broadleaf trees are found in amongst the beech, often in gullies and valleys.

Beech forest is very different in appearance from the other forests. The beech trees are usually of a fairly uniform height with a dense leaf canopy; a soft green light filters through their foliage to the mosses and ferns on the forest floor. Epiphytes and lianes are few, and the place of an understorey of trees is taken by beech saplings and a few species of small-leaved shrubs. Where beech grows on dry hillsides the forest floor may have only a covering of fallen leaves and patches of white or yellow-brown moss. Twiggy shrubs form a sparse undergrowth. In very wet regions beech trees are moss-clad and rise from a spongy carpet of moss and filmy ferns. Large clumps of mistletoe, either scarlet or yellow-flowered are a feature of beech forest in summer.

To make a study of native trees does not mean

an excursion to remote and untouched areas of forest. Throughout New Zealand there are many large and small patches and pockets of native vegetation where a number of different trees may be seen.

Native trees on farmland are often an indication of the forest that once covered the district. Such groves of trees are useful in providing shelter and shade for stock, at the same time giving a pleasant aspect to pasture land. In northern farmland puriri and taraire are handsome shade trees. Totara, kahikatea, beech, cabbage trees, manuka and kanuka are seen, in many districts. Karaka, ngaio and pohutukawa flourish near the coast, while in inland places beech, ribbonwood, and fuchsia are common.

Remnants of native bush are often found alongside rivers and streams, especially in steep-sided gorges. Kowhais and lacebarks are particularly beautiful along riverbanks and along the shores of lakes and inlets. In the north, pohutukawa is a tree of the shoreline, while in some southern areas rata is a coastal tree. Even on barren, windswept coasts, gullies running back from the shore may be a rewarding collecting ground for small trees.

Large areas of hill country and poor land are often covered with manuka scrub and the taller kanuka. In the shelter beneath manuka, seedlings of some forest trees flourish and soon appear above the manuka. In the north, kauri, tanekaha, and rewarewa are conspicuous on manuka country where there once was forest. Other small, quick-growing trees that are early members of a "second growth" bush are mahoe, wineberry, rangiora, five-finger, and karamu.

NOTES ON TEXT AND PLATES

The tree heights are given in metres (1 metre=39.37 inches). For an approximate conversion to feet, multiply by $3\frac{1}{4}$. Other measurements are given in centimetres and millimetres, and a measuring-rule is printed inside the back cover.

The trees appear in the same order in which they may be found in standard, recent botanical reference books—a natural order of classification based upon their relationship to each other. Latin and plant family names as well as common names are given.

To use the Latin name is the only certain way of referring to any plant, as common names are often shared by several plants, sometimes quite dissimilar in character. Some plants have several common names which vary from district to district. The Latin names are international; any botanist would immediately recognise the plant described when he knows its Latin generic and specific names which together may be given to no other plant in the world.

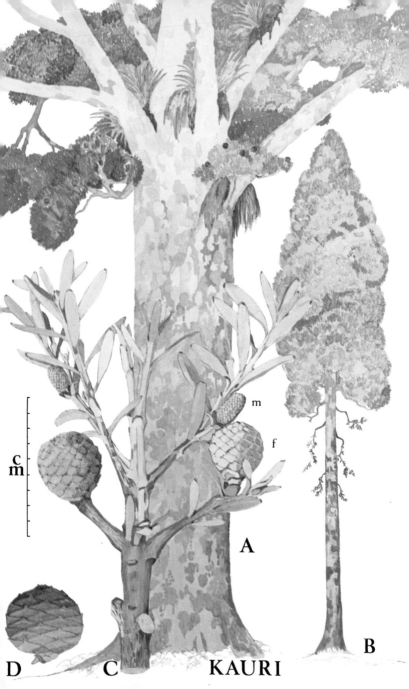

m

f

c
m

A

B

D C KAURI

KAURI

Agathis australis Salisb.

Family: *ARAUCARIACEAE*

KAURI is massive forest tree reaching 30m (100 ft) or more with a straight, cylindrical, unbranched trunk of great diameter in mature trees. The bark is light grey, hammer-marked, and comes away in flakes. Young trees have straight, tapering trunks with conical crowns unlike the dome-like crowns of old trees with their huge spreading branches.

The foliage of kauri saplings is often bronze with the leaves longer and more pointed than those of adult trees that are short, stiff, and blunt-ended. Both types of foliage may be seen on young trees. Male and female flowers are borne on the same tree on adjacent branchlets.

The mature cones are globular, dark green and resinous, and when ripe fall to pieces and scatter the winged seeds.

Kauri trees occur naturally only in the North Island, north of Maketu on the east coast and Kawhia on the west. They are widely planted in parks and gardens in the North Island and in some South Island localities.

PLATE 1   KAURI

    A Mature tree

    B Young tree

    C Adult foliage with male (m) and female (f) cones

    D Ripening cone

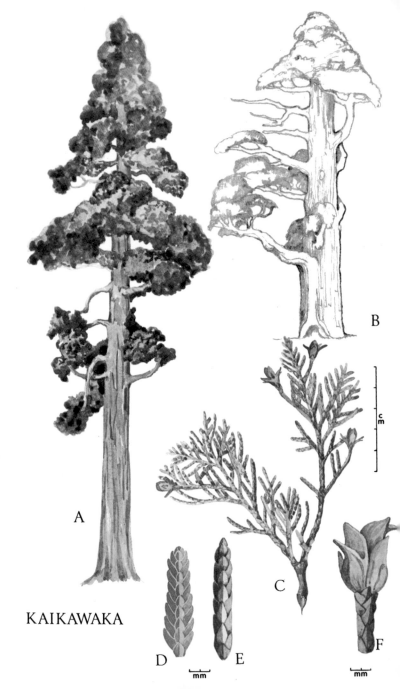

KAIKAWAKA

KAIKAWAKA

Libocedrus bidwillii Hook. f.

Family: *CUPRESSACEAE*

KAIKAWAKA or pahautea, New Zealand cedar. Kaikawaka is a small tree reaching 16m (50 ft). At high altitudes it is often much shorter, very stunted, gnarled or contorted by the wind. Such trees may be very old. A young tree growing in a sheltered place is very symmetrical and has a perfectly straight tapering trunk with pinkish-tan bark that peels off in narrow strips.

The foliage is dull blue-green. Juvenile foliage differs from that of the adult, the branchlets are flattened and feathery, while adult branchlets are markedly four-angled. Male and female flowers are very small. The mature cones are small and woody.

Kaikawaka is found in both the North and South Islands, south of Mt Te Aroha and Mt Egmont to Foveaux Strait. It is a tree of high altitude forests and is commonly found near the upper limit of the bush on mountainsides.

PLATE 2   KAIKAWAKA

    A   Young tree
    B   Ancient tree
    C   Adult foliage with cones
    D   Juvenile foliage, magnified
    E   Adult foliage, magnified
    F   Cone, magnified

A

B

C

cm

TOTARA

D

m
m

TOTARA

Podocarpus totara G. Benn. ex Don

Family: *PODOCARPACEAE*

TOTARA is a large forest tree up to 30m (100 ft) high with very dense, dark green foliage not unlike the English yew. The trunk and lower limbs are stout with rough reddish-brown bark that comes off in strips.

The stiff leaves are closely set on the branchlets. Totara saplings have brownish leaves that are larger and set further apart than those of the adult trees. Male and female flowers are borne on separate trees. The fruit is set on a red, berry-like receptacle.

Totara is found in both the North and South Islands in lowland forests, sometimes forming almost pure stands. The very similar thinbarked or Hall's totara (*Podocarpus hallii*) grows at higher altitudes.

PLATE 3   TOTARA

      A   Mature tree
      B   Juvenile foliage
      C   Adult foliage with ripe fruit
      D   Enlargement of fruit

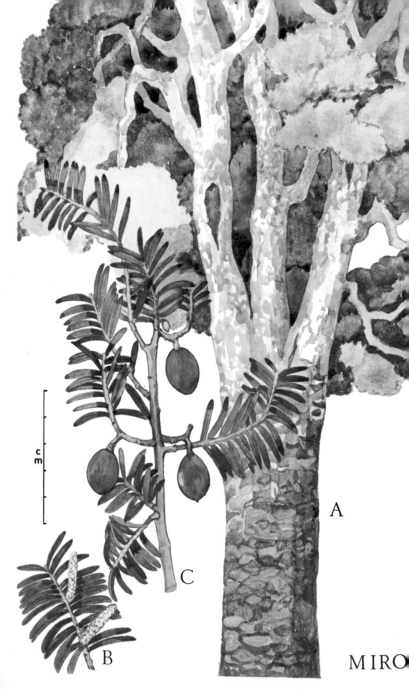

A

B

C

c
m

MIRO

MIRO

Podocarpus ferrugineus G. Benn. ex Don

Family: *PODOCARPACEAE*

MIRO is a forest tree reaching 25m (80ft) with dark green yew-like foliage. The trunk is smooth, grey and with hammer-marked, flaking bark.

The leaves are set in two rows on the branchlets and each leaf is curved. Older leaves are dark but young sprays of foliage are a bright fresh green. Male and female flowers are borne on separate trees. The fruit is plumlike with a waxy bloom over the clear purplish pink flesh. Both fruit and foliage are strongly aromatic.

Miro is found in forests throughout New Zealand.

PLATE 4   MIRO

    A   An old tree
    B   Male flowers (stroboli)
    C   Ripe fruits

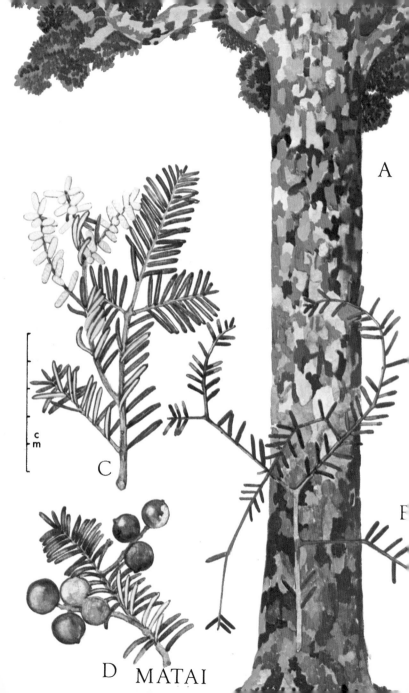

A

C

c
m

D MATAI

MATAI

Podocarpus spicatus R. Br. ex Mirbel

Family: *PODOCARPACEAE*

MATAI or black pine is a forest tree reaching 30m (100 ft) with a tall, straight trunk. The flaking, dark grey bark is conspicuously hammer-marked.

The foliage of adult trees is dark blue-green, with the leaves slightly curved and whitish beneath. The leaves are aromatic when crushed. The saplings have long wiry angular branches with brownish green leaves set wide apart. Male and female flowers are borne on separate trees. The fruits are globular, blue-black, and fleshy, on short, upright, branchlets.

Matai is found throughout New Zealand in lowland forests.

PLATE 5   MATAI

    A   Trunk

    B   Juvenile foliage

    C   Adult foliage and male ''flowers''

    D   Ripe fruits

A

B

C

c
m

D

RIMU

mm

RIMU

Dacrydium cupressinum Lamb.

Family: *PODOCARPACEAE*

RIMU or red pine is a lofty forest tree that grows to a height of 50m (165 ft) overtopping the surrounding forest canopy. The trunk is straight and tapering with dark flaking bark. The branchlets hang downwards giving the crown a distinctive "weeping" appearance. This weeping habit is more pronounced in the young trees that have longer, softer branchlets.

The leaves are small, close-set, and harsh to the touch. and the foliage has a rich olive-green appearance when viewed from a distance. Male and female flowers are on separate trees. The fruit is a dark nut set on a fleshy scarlet base.

Rimu is found in forests throughout New Zealand.

PLATE 6   RIMU

      A   Mature tree

      B   Young tree

      C   Foliage with ripe fruit

      D   Enlargement of fruit and leaves

A

C

F

E

D

B

KAHIKATEA

KAHIKATEA

Podocarpus dacrydioides A. Rich.

Family: *PODOCARPACEAE*

KAHIKATEA or white pine is a very tall tree with a grey tapering trunk. Old trees grow to a height of 50m (165 ft) or more and have rather sparse foliage high on the branches. Young trees are symmetrical with dense foliage.

The leaves of kahikatea are small and scale-like on adult trees close-set on weeping branchlets. Saplings have a bronze, feathery juvenile foliage of narrow, curved leaves. Male and female flowers are borne on separate trees. The fruits, often set in great quantities, are bluish-black nuts on a fleshy orange-red base.

Kahikatea is found throughout New Zealand in lowland forest and in dense stands on swampy land. Groves of young kahikatea are often found where farmland has been brought in from drained swamps.

PLATE 7   KAHIKATEA

A   Stand of tall, old trees
B   Young trees
C   Juvenile foliage
D   Adult foliage with male ''flowers''
E   Fruiting blanchlet
F   Enlargement of fruit

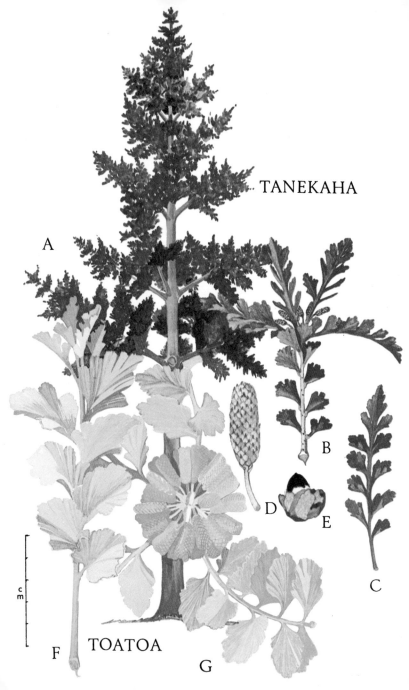

TANEKAHA

A

B

C

D

E

TOATOA

F

G

TANEKAHA AND TOATOA

Phyllocladus trichomanoides Don

Family: *PODOCARPACEAE*

TANEKAHA and TOATOA, the celery pines. Tanekaha is a forest tree that grows to 20m (65ft) with a straight tapering trunk from which the branches grow in a series of whorls. The bark is light grey and smooth. Young trees are very symmetrical with a "Christmas tree" outline and are often reddish-bronze in colour.

The foliage of mature trees is deep green. The "leaves" (cladodes) are flattened lobes of the branchlets, not true leaves, and are arranged in whorls about the side-branches. The male flowers are in terminal clusters; the female are set on the margins of modified cladodes.

Tanekaha grows in the north of the North Island and in the north and north-west of the South Island to Westport.

TOATOA (*Phyllocladus glaucus* Carr.) is in the northern forests only; it is a small tree that has bluish-green foliage, larger thicker cladodes, and fruits in rounded clusters.

PLATE 8   TANEKAHA and TOATOA

    A A young tanekaha
    B Tanekaha foliage
    C Tanekaha cladodes
    D Tanekaha, male cone magnified
    E Tanekaha, fruit magnified
    F Toatoa foliage
    G Toatoa, cluster of male cones

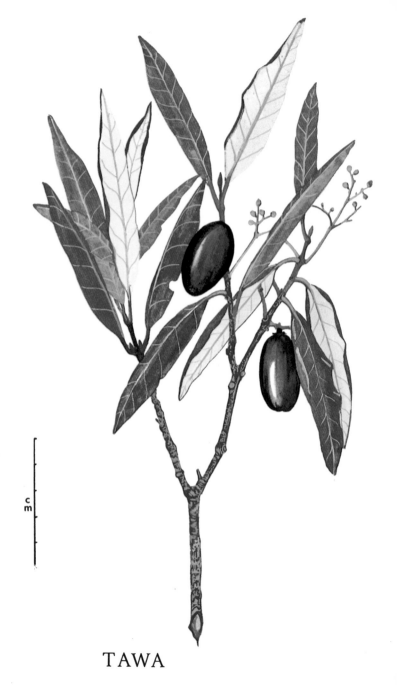

TAWA

TAWA

Beilschmiedia tawa (A. Cunn.) Benth.

Family: *LAURACEAE*

TAWA is a tall forest tree up to 25m (80ft) in height with a straight, smooth trunk. It forms pure stands in some localities.

The foliage is pale and rather lacy. The leaves are narrow and yellowish green with paler undersides. Often they lie crisp and undecayed on the forest floor. The flowers are small and borne in pale green sprays in early summer. They are followed by large blue-black fruits in late summer. Tawa is found throughout the North Island and in northern areas of the South Island.

PLATE 9   TAWA

Foliage and ripe fruit

mm

b

c

C

B

A

PIGEONWOOD

PIGEONWOOD

Hedycarya arborea J. R. and G. Forst.

Family: *MONIMIACEAE*

PIGEONWOOD is a small tree that grows to a height of about 15m (50 ft) with bark that is brown and fairly smooth.

The foliage is dark green and glossy. The leaves are rather thick, glossy on the upper surface but pale beneath with distantly and irregularly toothed margins. The flowers are small; male and female are borne on separate trees. The female flowers are followed by heavy clusters of shiny drupes that pass from dark green through yellow to deep orange as they ripen.

Pigeonwood is plentiful in forests from the north of the North Island to Banks Peninsula and Milford Sound.

PLATE 10   PIGEONWOOD

    A   Foliage and fruits

    B   Male flowers; b male flower magnified

    C   Female flowers; c female flower magnified

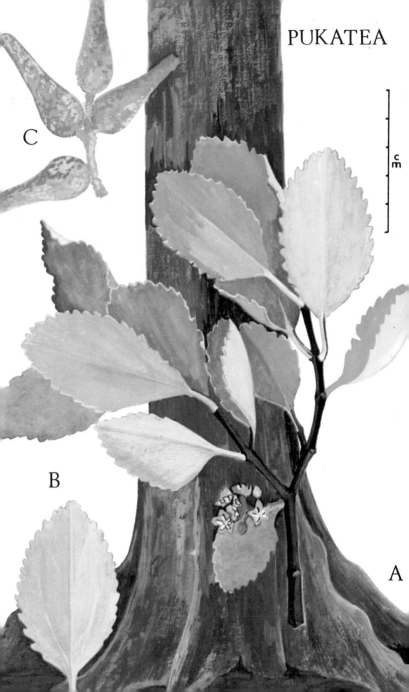

PUKATEA

C

B

A

cm

PUKATEA

Laurelia novae-zelandiae A. Cunn.

Family: *MONIMIACEAE*

PUKATEA is a tall spreading tree reaching 35m (114 ft) in height with a smooth dark grey bark. The base of the trunk develops widely flanged buttresses. Pukatea grows where the drainage is poor and in swampy forest. Young trees are symmetrical in outline and often have bright yellow-green foliage.

The leaves are shiny, bright green and have evenly toothed margins; the midrib and veins are inconspicuous. The branchlets are four-angled and often dark purple. Pukatea flowers are small and borne on short racemes. The fruits are flask-shaped and split to release the silky seeds.

Pukatea is found in the North Island and the north of the South Island.

PLATE 11   PUKATEA

    A   Pukatea trunk with plank buttresses
    B   Foliage
    C   Fruit

MAHOE

MAHOE

Melicytus ramiflorus J. R. and G. Forst.

Family: *VIOLACEAE*

MAHOE or whiteywood is a small tree up to 10m (33 ft) high with whitish bark and dense foliage.

The new leaves in the early summer are a conspicuous yellowish green. The mature leaves are dark green and thick, with prominent veins and toothed margins. Decaying mahoe leaves frequently form perfect "skeleton" leaves. The flowers are small, greenish yellow and arranged in little bunches on the stems below the leaves. Male and female flowers are borne on separate plants. The fruits are violet when fully ripe.

Mahoe is particularly common in second-growth bush and in coastal forest.

PLATE 12   MAHOE

    A   Foliage and ripe fruits
    B   Flowers (female)
    C   Male flower, magnified
    D   Female flower, magnified

C   REWAREWA   B

A

REWAREWA

Knightia excelsa R. Br.

Family: *PROTEAEAE*

REWAREWA, or New Zealand honeysuckle, is a forest tree reaching 30m (100 ft), rather poplar-like in habit. It is particularly noticeable in regenerating bush where it stands well above the surrounding lower-growing trees.

The foliage is dull green sometimes with a rusty appearance. Sapling rewarewas have very long thin pale green leaves with sharply-toothed margins. The adult leaves are shorter and thicker and stand stiffly upright at the ends of the branchlets. The flowers are borne in compact clusters on small branchlets below the leaves. They are long and slender in bud, dark red and velvety. The petals curl back when the bud opens. The fruits are clusters of dry, rusty red "pods" that split to release the seeds.

Rewarewa is found in the North Island and in the Marlborough Sounds.

PLATE 13   REWAREWA

    A   Foliage and flowers

    B   Leaves: *left*, adult; *right*, juvenile

    C   Fruits

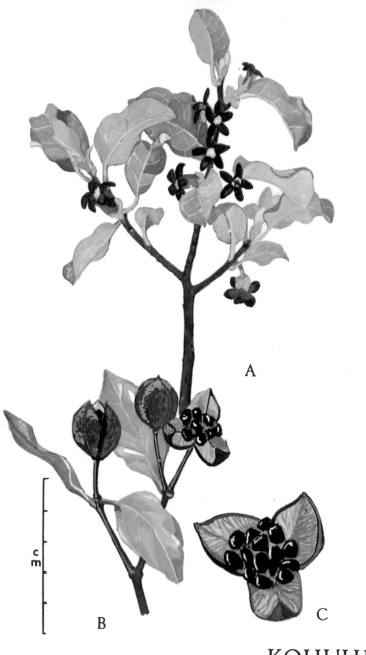

c
m

A

B

C

KOHUHU

KOHUHU

Pittosporum tenuifolium Sol. ex Gaertn.   .

Family: *PITTOSPORACEAE*

KOHUHU is a small tree up to 10m (33 ft) in height with smooth blackish bark. The light silvery-green leaves are shiny with wavy margins. The flowers are small, very dark red—almost black, and are borne singly in the axils of the leaves. They are followed by seed capsules that split when ripe to show the black seeds set in a sticky gum. Kohuhu and other pittosporums are often distributed by the gummy seeds adhering to the feathers, beaks, or feet of birds.

Kohuhu is a tree of shrubland and light bush and is common throughout New Zealand except in Westland. Several forms, especially those with coloured leaves, are widely grown in gardens.

PLATE 14   KOHUHU

      A   Foliage and flowers

      B   Fruits

      C   Capsule showing seeds

MANUKA

MANUKA

Leptospermum scoparium J. R. and G. Forst.

Family: *MYRTACEAE*

MANUKA or tea-tree one of the most widespread of native plants, is a slender tree growing to a height of 8m (26 ft), often in dense stands of individuals about the same height. The slender trunks are covered with grey bark that comes away in strips.

The foliage is grey-green to bronze-green. The leaves are small and close-set with sharply pointed tips. The flowers are white petalled with a reddish central disc. Occasionally pink and red-flowered manukas are found as wild plants. The hard seed capsules range from bronze to woody grey as they mature and when ripe they split to release the fine "sawdust" like seeds.

Manuka is often considered to be a weed of hilly farmland nevertheless it also gives a much needed protective cover on large areas of land with unstable, eroding soils. In many places it remains stunted and rarely attains tree size.

PLATE 15   MANUKA

    A   Windswept manuka
    B   Flowering branchlets
    C   Flower, magnified
    D   Capsule, magnified

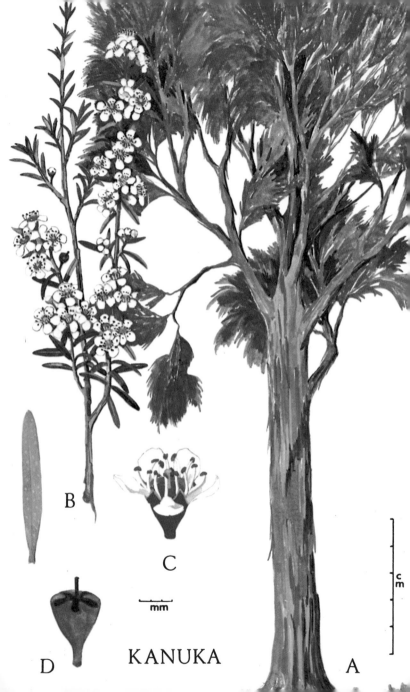

B

C

D

KANUKA

A

KANUKA

Leptospermum ericoides A. Rich.

Family: *MYRTACEAE*

KANUKA is the taller growing tea-tree, reaching a height of 15m (50 ft) sometimes with a trunk several feet in diameter in very old specimens. The bark is reddish and peels off in strips.

The foliage is similar in appearance to that of manuka but kanuka often has a graceful weeping appearance when growing in a sheltered place. The leaves are softer and not as sharply pointed. Like manuka, kanuka also has an abundance of flowers readily distinguished at a distance from the whiter manuka by their pinkish-brown tinge. The flowers are smaller and borne in little bunches (fascicles) along the stems. The small seed capsules split to disperse the fine seeds.

Kanuka is found all over New Zealand and may often be seen along stream banks and as a small shelter tree on farmland.

PLATE 16   KANUKA

  A   Kanuka tree
  B   Flowering branchlet
  C   Flower, magnified
  D   Capsule, magnified

A

B

C

D

NORTHERN RATA

NORTHERN RATA

Metrosideros robusta A. Cunn.

Family: *MYRTACEAE*

NORTHERN RATA is a lofty forest tree mostly overtopping the surrounding trees. It grows to a height of 30m (100 ft). The northern rata usually begins life as a perching seedling high in the branches of other forest trees. Roots descend from the young rata down the trunk of the host (often a rimu), to reach the ground. In time these roots become thick and woody eventually uniting to form a massive trunk inside which is the decaying trunk of the host—ancient rata trees are often hollow.

The foliage of rata is dark green, the leaves are small and leathery with rounded indented tips and dotted with glands on the undersides. The flowers are a mass of bright red stamens borne in sprays on the tips of the branches. The small capsules split to release the fine, woody seeds.

Northern rata is found in forest in the North Island and in the northern areas of the South Island to Westland.

PLATE 17   NORTHERN RATA

      A   A rata trunk showing descending and encircling roots

      B   A spray of flowers

      C   Leaves

      D   Capsule

A

B

C

SOUTHERN RATA

Metrosideros umbellata Cav.

Family: *MYRTACEAE*

SOUTHERN RATA is a smaller tree than the northern rata, and grows up to 20m (65 ft) high. The bark is light grey and papery on the trunk and branches. Old trees in exposed situations are very gnarled, and in mountain areas southern rata may be a small shrub flowering as brilliantly as the larger trees. Both ratas flower in midsummer and may in some years, bear masses of flowers and in others, almost none. The leaves of southern rata are dark green and shiny with long tapering tips. The fruit is a woody capsule.

Southern rata grows south of Northland (in certain localities only) and in quantity in South Island forests, especially west of the Main Divide.

PLATE 18   SOUTHERN RATA

    A   A mature tree

    B   A flowering branchlet

    C   Capsule

A   B   C   D   E

POHUTUKAWA

POHUTUKAWA

Metrosideros excelsa Sol. ex Gaertn.

Family: *MYRTACEAE*

POHUTUKAWA, the New Zealand "Christmas tree" is a large spreading tree up to 25m (80ft) high with gnarled trunks and branches. The bark is dark grey and grooved; the lower limbs are often festooned with tufts of reddish aerial roots.

Pohutukawa has dark green foliage with a silvery white appearance when the trees are covered with buds, new growth, or young seed capsules. The leaves have a smooth upper surface and are whitefelted beneath; the leaf buds and young shoots are also white. The flowers are produced in great abundance about the middle of December and their rich crimson colour comes from the mass of stamens, as the petals are small and fall soon after the buds open. The capsules following the flowers are white and when they split large quantities of small thin seeds accumulate on the ground beneath the trees.

Pohutukawa is a tree of northern coastal places where it often grows on steep cliffs just above the water or as a fringe around beaches and lakes. It is widely cultivated in areas south of its natural range which is from North Cape to Poverty Bay and Taranaki.

PLATE 19   POHUTUKAWA
     A   An old tree
     B   Spray of flowers
     C   Silvery leaf (underside)
     D   Flowers
     E   Seed capsules

HINAU

HINAU

Elaeocarpus dentatus (J. R. and G. Forst.) Vahl.

Family: *ELAEOCARPACEAE*

HINAU is a forest tree reaching 20m (65 ft) with smooth bark and dull green foliage.

The leaves are deep green above, pale and silky below with toothed, rolled-back margins. There are conspicuous domatia at the angles of the veins and midrib. The flowers of hinau are creamy-white with fringed petals borne in profusion on long racemes. They are followed by the rosy fruits, several to a stem.

Hinau is plentiful in forests from North Cape to Foveaux Strait.

The closely related pokaka, (*Elaeocarpus hookerianus*) is also a forest tree. It differs from hinau in having coarsely toothed leaves, greenish sprays of small flowers and a very distinctive juvenile form.

PLATE 20   HINAU

A   Flowering branch, note pits (domatia) on undersides of leaves

B   Enlarged flower

C   Ripe fruits

B

A

C

WINEBERRY

WINEBERRY

Aristotelia serrata
(J. R. and G. Forst.) W. R. B. Oliver

Family: *ELAEOCARPACEAE*

WINEBERRY or makomako is a small tree up to 10m (33 ft) tall with rather shiny reddish bark and light open foliage often with a pink tinge.

The leaves are light green with deeply and sharply-toothed margins. The undersides of the leaves are sometimes purplish-pink. The flowers are small and rosepink borne in great profusion on the branches just below the leaves. Male and female flowers are found on separate trees. They are followed by the dark red or black berries.

Wineberry is found throughout New Zealand, especially along roadsides in bush country and in second-growth bush areas.

PLATE 21   WINEBERRY

    A   Flowering branch
    B   Male flower magnified
    C   Fruits

A

B

C

HOUHERE

cm

HOUHERE

Hoheria populnea A. Cunn.

Family: *MALVACEAE*

HOUHERE. Lacebark is the name given to several species of *Hoheria*, all small, graceful trees with white flowers. The name "lacebark" refers to a netted fibrous layer found beneath the outer bark. The lacebark illustrated is one often cultivated as an ornamental tree.

The leaves are light green and sharply toothed. The flowers are white and followed by the green, winged fruit.

Hoheria populnea is found wild only as far south as the Waikato but other lacebarks occur throughout New Zealand especially in forest clearings and along streambanks. They all bear masses of white flowers in late summer.

PLATE 22   HOUHERE

     A   Flowering branch

     B   Adult and (left) juvenile leaf

     C   Fruits

KAMAHI

KAMAHI

Weinmannia racemosa Linn. f.

Family: *CUNONIACEAE*

KAMAHI is a tall spreading tree growing to a height of 26m (85 ft). The trunk and lower limbs are heavy and covered with a grey bark.

The foliage is a dark brownish-green on older trees. Young kamahi saplings are of a much richer colour, ranging from bronze to crimson with smaller, several-foliate leaves. The leaves of adult trees are leathery with coarsely toothed margins and dark veins on the underside. Kamahi flowers profusely in the early summer. The trees are a mass of upright flower stalks each with many small fluffy flowers pinkish cream in colour. They are followed by the seed capsules which, as they mature, tint the trees a deep rusty-red.

Kamahi is an abundant forest tree from the Coromandel Peninsula to Stewart Island. In the north the similar towai (*Weinmannia silvicola*) is widespread.

PLATE 23   KAMAHI

A   Kamahi showing large, spreading branches
B   Foliage and flowers
C   Juvenile leaves
D   Fruits

A

C

B

KOWHAI

KOWHAI

Sophora microphylla Ait.

Family: *PAPILIONACEAE*

KOWHAI. The two species of kowhai tree are very similar in habit flower, and fruit. Both grow to a height of about 12m (40 ft).

The difference lies in the leaves—one has small, roundish leaflets, many to a stem (S. *microphylla*); the other has larger more elongated leaflets fewer to a stem (S. *tetraptera*). There are many intermediate forms. Both species are more or less without leaves by the end of the winter, and the clusters of golden-yellow flowers are borne on the bare branches in the late spring. The new leaves come with the flowers and the long, grey pods develop. These become papery and torn to release the hard yellow seeds.

Kowhai is found at the edge of the forests, on stream banks and along the shores of lakes and inlets. S. *tetraptera* is found in the North Island; S. *microphylla* from North Cape to Southland. (illustration)

PLATE 24   KOWHAI

    A Flowering tree
    B Spray of flowers showing small leaflets
    C Seed pod

c
m

A

B

C

D

E

BLACK BEECH

BLACK BEECH AND MOUNTAIN BEECH

Nothofagus solandri (Hook. f.) Oerst.

Family: *FAGACEAE*

BLACK BEECH grows to a height of 25m (80 ft) and has a rough, dark trunk.

The foliage is deep, dull green, and dense. The leaves are small with plain margins very slightly rolled back; they are dark green on the upper surface and pale beneath. Like those of the other native beeches, the flowers are small, but the male flowers have scarlet stamens and are borne in such profusion that the whole tree is bright for a short time in early summer. The fruit is a small woody cupule.

MOUNTAIN BEECH (*Nothofagus solandri* var. *cliffortioides* (Hook. f.) Poole) is a smaller tree than black beech with more pointed leaves often with very markedly rolled leaf margins on young growth. The flowers and fruits are similar to those of black beech.

Black and mountain beech both grow from the centre of the North Island southwards, with the latter growing higher on the mountains and extending further south to reach Foveaux Strait.

PLATE 25   BEECH
   A   Black beech
   B   Black beech, juvenile foliage
   C   Black beech, foliage
   D   Mountain beech, foliage
   E   Male flower of mountain beech

B

c
m

C

D

SILVER BEECH

SILVER BEECH

Nothofagus menziesii (Hook. f.) Oerst

Family: *FAGACEAE*

SILVER BEECH is a tall and beautiful tree reaching a height of 30m (100 ft). At the bush line in mountain areas silver beech may form an ''elfin forest'' of low, gnarled trees hung with moss and lichen. The trunks and limbs of large trees are thick and covered with rough grey bark.

The foliage is light green and arranged in layers on the horizontally spreading branches. Silver beech leaves are thick with toothed margins; the teeth are blunt and double in this species. The flowers are small and borne on the spring shoots—female flowers obscurely placed within scales, the male flowers more obvious because of their orange-pink or yellow anthers. The fruit is small and woody (a cupule).

Silver beech is found from near Auckland south to Foveaux Strait, but not on Mt Egmont, where no beech species are found.

PLATE 26   SILVER BEECH

   A   A large silver beech

   B   Foliage and a leaf showing double-toothed margins

   C   Male flower with bright stamens

   D   Fruit—a cupule with one of the nuts from within it

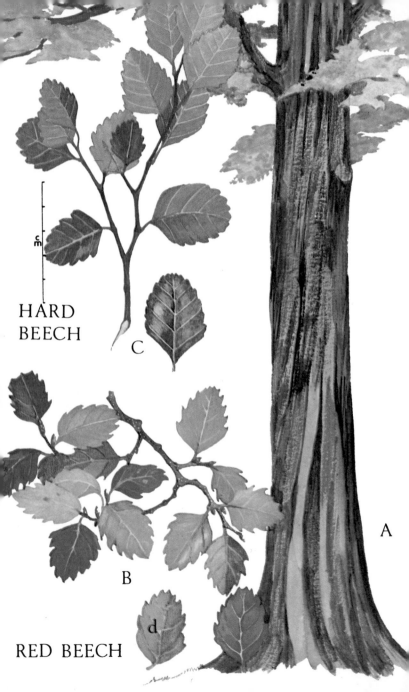

HARD
BEECH

C

RED BEECH

B

d

A

RED BEECH

Nothofagus fusca (Hook. f.) Oerst.

Family: *FAGACEAE*

RED BEECH is a forest tree up to 30m (100 ft) high with a straight fluted trunk. The bark is dark and furrowed. When freshly cut the wood is bright crimson.

The foliage is light green and appears almost deciduous as brightly tinted leaves fall in quantity. The leaves are sharply toothed with small pits (domatia) in the angle of the veins and midrib.

Hard beech, *Nothofagus truncata* (Col.) Ckn., is also a tall tree resembling red beech. Its leaves have rounded teeth and no domatia.

Both these beeches have small flowers and woody fruits. They are found in lowland and hill country forest in the North Island. Hard beech is only in the north of the South Island, whilst red beech grows as far south as Fiordland.

PLATE 27   RED BEECH

    A   A red beech tree
    B   Red beech foliage and leaves
        (domatia marked "d")
    C   Hard beech foliage

A

B

C

KARAKA

KARAKA

Corynocarpus laevigatus J. R. and G. Forst.

Family: *CORYNOCARPACEAE*

KARAKA is a large tree of coastal situations and coastal forest. It has dense foliage and grows to a height of 20m (65 ft). The trunk is smooth and grey. Trees growing in exposed places are windshorn into a dense low thicket with a thick leaf canopy above the close-standing trunks.

Karaka has very dark green, large, glossy leaves with smooth margins. The flowers are small and greenish; they are followed by large clusters of glossy green fruits that ripen through yellow to orange.

Karaka is found near the coast from North Cape to Banks Peninsula.

PLATE 28   KARAKA

A   Young trees
B   Foliage and fruit
C   Coastal karaka forest

A

B

TITOKI

TITOKI

Alectryon excelsus Gaertn.

Family: *SAPINDACEAE*

TITOKI grows to a height of 12m (40 ft) and has a smooth dark trunk and rather dull brownish green foliage.

The leaves are olive green set in pairs on a felty brown leaf-stalk. The buds and young shoots have also a brown velvety surface. The flowers are small in loose sprays and are furry on the outside and dull purple within. The fruits are velvety and split to reveal a shiny black seed surrounded by a scarlet fleshy aril.

Titoki is a handsome tree found throughout the North Island to Banks Peninsula and Westland in the South Island. It is often seen as a shelter tree on farmland.

PLATE 29   TITOKI

    A   Leaf

    B   Cluster of fruits

KOHEKOHE

c
m

B   A   C

KOHEKOHE

Dysoxylum spectabile (Forst. f.) Hook. f.

Family: *MELIACEAE*

KOHEKOHE is a dense-foliaged tree that grows to a height of 17m (55 ft). In coastal forest it is often much shorter with a thick, windshorn canopy.

The foliage is dark green except for the young growth, which is a bright yellow-green The large, glossy leaflets are set in several pairs on a long leaf stalk. The sprays of waxy flowers are unusual in that they spring from the bark of the trunk or larger branches well below the foliage. They open in the autumn or early winter. The hanging bunches of fruits that follow split to expose the seeds with their fleshy red covering.

Kohekohe is found in the North Island and in the Marlborough Sounds.

PLATE 30   KOHEKOHE

     A   Leaf

     B   Flower spray

     C   Fruits

A

B

FIVE FINGER

FIVE-FINGER

Pseudopanax arboreum (Murr.) Philipson

Family: *ARALIACEAE*

FIVE-FINGER is a small tree about 8m (26 ft) high with a smooth, slender, branching trunk, purplish stems, and dark foliage.

The leaves are normally divided into five stalked-leaflets, dark green above and paler below. The margins are coarsely toothed. Five-finger flowers profusely, male and female in large rounded trusses on separate plants. The purple inflorescences are followed by the flattened dark fruits on the female trees.

Orihou, (*P. colensoi*) is similar to the five-finger although it often has three leaflets. Orihou leaflets have no stalk.

Five-finger is plentiful in lowland bush and second growth forest throughout New Zealand, while orihou is a common tree or shrub at higher elevations from Little Barrier to Stewart Island.

PLATE 31   FIVE-FINGER

    A   Flowering branch

    B   Fruit

c
m

A

B

C

D

PUKA

PUKA AND PAPAUMA

Griselinia lucida Forst. f.

Family: *CORNACEAE*

PUKA or broadleaf is a shrub or small tree up to 8m (26ft) high. It is frequently an epiphyte (a perching plant) on other trees. The puka has a rounded crown of glossy rich green foliage and fluted, grey roots descend to the ground.

The leaves are leathery, very shiny, pale on the underside, and markedly asymmetrical. The flowers are small and greenish; male and female are borne on separate trees. Female plants bear sprays of purple-black berries.

Puka is found from North Cape to Foveaux Strait.

PAPAUMA, (*Griselinia littoralis* Raoul), is the more common broadleaf, a tree growing to a height of 17m.

Its flowers and berries are like those of puka.

Papauma grows from North Auckland to Stewart Island and is a frequent small tree in subalpine forest.

PLATE 32   PUKA

A   The descending roots of an epiphytic puka

B   Puka foliage and fruit

C   Puka leaf showing unequal base

D   Papauma leaf, more oval and less unequal at base.

A

B   C

KARAMU

KARAMU

Coprosma lucida Raoul

Family: *RUBIACEAE*

KARAMU is a shrub or small tree up to 5m (16 ft) high with dark green glossy foliage and pale grey smooth bark.
. The leaves are thick and leathery, dark green and shiny above, and pale below. Male and female flowers are borne on separate trees in many-flowered clusters. The greenish flowers have large stamens and anthers as they are wind pollinated. The fruits are bright orange and shiny.

Karamu is one of the most widely distributed of several large-leaved, orange-fruited coprosmas. It is abundant in bush and shrubland, in all parts of the country.

PLATE 33   KARAMU

A   Fruiting branch
B   Female flower showing large stigmas
C   Male flower showing pendulous stamens

B   NGAIO   A

NGAIO

Myoporum laetum Forst. f.

Family: *MYOPORACEAE*

NGAIO is a tree of coastal places reaching a height of 8m (26 ft) with heavy spreading branches covered with grey corky bark.

The foliage is a distinctive light olive green. The leaves are thin and shiny covered with pale dots with a few teeth on the margins towards the tip. The leaf buds are blackish and sticky, a feature that distinguishes the native ngaio from the toothed-leaved Tasmanian species that is widely grown as a shelter tree in seaside localities. The flowers are small, cream and dotted with brown; the berries are magenta when ripe.

Ngaio is found in the North Island and as far south as Otago.

PLATE 34   NGAIO

A   A tree showing the spreading branches
B   Foliage and fruit

A

B

C

D

E

MANAWA

cm

MANAWA

Avicennia resinifera Forst. f.

Family: *VERBENACEAE*

MANAWA or mangrove is a small tree that grows to height of about 8m (25 ft) although it is more often much shorter. Like its tropical relatives, mangrove grows in tidal mud about our northern coasts where it is surrounded by water at high tide. At low water the upright "breathing" roots may by seen standing above the mud. Mangrove trunks are hoary and gray. Old trees have rather twisted and gnarled branches.

The foliage is a rich olive green; the leaves are leathery with prominent veins. The small flowers are followed by masses of yellowish-buff, velvety-coated fruits. These fruits fall with well developed cotyledons and very quickly develop a short stem and roots when they are washed up by the receding tide. The various stages of the developing seedlings may be found at high tide mark near mangrove swamps.

Mangrove is found only in the North Island, north of Kawhia and Opotiki. It is very abundant where conditions are suitable for its growth.

PLATE 35   MANAWA

    A   An old tree with exposed breathing roots
    B   Foliage with mature fruits
    C   Seeds when shed
    D   "Purse" stage ready to take root
    E   Seeding from high tide mark